Nougbodjingni Damien Makponse

Agricultural farming practices in the face of climate change

Nougbodjingni Damien Makponse

Agricultural farming practices in the face of climate change

The land of land pacts (Savalou-Bantè)

ScienciaScripts

Imprint

Any brand names and product names mentioned in this book are subject to trademark, brand or patent protection and are trademarks or registered trademarks of their respective holders. The use of brand names, product names, common names, trade names, product descriptions etc. even without a particular marking in this work is in no way to be construed to mean that such names may be regarded as unrestricted in respect of trademark and brand protection legislation and could thus be used by anyone.

Cover image: www.ingimage.com

This book is a translation from the original published under ISBN 978-613-8-42334-8.

Publisher:
Sciencia Scripts
is a trademark of
Dodo Books Indian Ocean Ltd. and OmniScriptum S.R.L publishing group

120 High Road, East Finchley, London, N2 9ED, United Kingdom
Str. Armeneasca 28/1, office 1, Chisinau MD-2012, Republic of Moldova, Europe
Printed at: see last page
ISBN: 978-620-3-01657-4

TABLE OF CONTENTS:

Acknowledgements

The accomplishment of this work was only possible thanks to the active participation of several people to whom I express my deep gratitude. I would particularly like to thank Professors Moussa GIBIGAYE (Senior Lecturer) and Brice TENTE (Full Professor) who guided me in the choice of this subject and who voluntarily agreed to supervise this dissertation despite their multiple occupations and to respond each time to our concerns. I would also like to thank the members of the jury for their presence despite their multiple occupations. They have my deepest gratitude.

I would also like to thank our teachers who have trained us from primary school through to university, particularly those at DGAT who have succeeded in making us the people we are today. I would also like to thank all the staff at DGAT.

To Mr Joël D. YALLOU, Head of the Cartographic and Communication Unit at LARES. May God bless you.

Summary

In recent years, the land pacts region has experienced the vulnerability and harmful effects of climate change. Communities are unanimous on the fact that climate risks have become very noticeable and are disrupting farming activities in particular. This study analyses the adaptation strategies used by farmers to improve their living conditions. The methodology used consists of collecting, processing and analysing results based on 120 farmers in two districts of each commune. The field results reveal that the indicators are: the delay in the start of the rains, the appearance of pockets of drought during the rainy season, the early end of the rainy season, the increase in daily temperature, the appearance of violent winds, the multiplication of pockets of drought during the rainy season, the increase in the level of plant attack and finally the poor spatial distribution of the rains. Added to all this are the main climatic risks: pockets of drought (83%), poor spatial distribution of rainfall (71%) and rising temperatures (39%). Appropriate methods of adaptation to climate change are needed.

Keywords: Climate risks; land pacts; adaptation strategies.

Introduction

According to the APRM's Agricultural Sector Performance Report, the agricultural sector plays an important role in the national economy. It provides 70% of jobs, and Benin's economy remains that of an underdeveloped economy (APRM, 2014). To do this, agriculture requires a large area to be sown for production and exploitation. It is in this sense that one of the specific characteristics of agricultural activity lies in the increased use of land, soil or land capital according to the most common denominations (Saliou, 2008). The production system is faced with numerous difficulties and requires efforts to ensure its sustainable development. Development can only be sustainable if hunger and malnutrition are eradicated (FAO, 2012). Yet these scourges are largely linked to climatic hazards. The effects of climate change (reduced agricultural production, deteriorating food security, increased incidence of floods and drought, spread of disease and increased risk of conflict due to the scarcity of land and water) are already evident (Montcho, 2014). In this global context, Africa, and sub-Saharan Africa in particular, appear to be the regions of the world most at risk. Sub-Saharan Africa's great vulnerability to climate change is due to its heavy dependence on agriculture and its limited capacity to adapt, which is due to a lack of resources and technologies (Hamani, 2007). The land pact territory, which is a development area in Benin, is already facing the effects of climate change, hence the need to find ways of adapting to better cope with climate change. This territory is therefore as much a space for consultation and sharing of representations around a common vision and identity as it is a simple space for carrying out (or receiving) economic projects. It is through this space for concerted action that the resilience of the area is strengthened, which can be structured around three interrelated axes: local economic diversification, the quest for local social cohesion and ecological viability (Raufflet, 2014). Local development is therefore a requirement for sustainable development. Cartography, being a science, can only be materialised by a map, which is therefore the language best suited to the restitution of territorial information, whether qualitative or quantitative. As a result, the scientific community is showing renewed interest in the uses of cartography in areas of public action (Martinais, 2008). It is in this context that the theme of **'mapping agricultural production practices in the face of climate change in the land pact area'** comes into play.

This work is divided into four chapters:

- The first chapter presents the theoretical framework and methodology;
- The second chapter presents the study area;
- The third chapter presents agricultural production as an indicator of climate change in the area covered by the land pacts;
- The fourth is entitled Risks and methods of adaptation to climate change.

CHAPTER 1

THEORETICAL FRAMEWORK AND METHODOLOGY

This chapter covers the literature review (state of the art), the problem, the hypotheses and research objectives, and the clarification of concepts.

1.1. Theoretical framework

1.1.1. Issues

In human history, the need to understand climate change has never been more urgent or important than in the 21ème century, especially in tropical areas where deforestation and species extinction are relatively more significant and living conditions more precarious (Bush and Flenley, 2007). These climate changes are reflected in global temperatures, which have risen by an average of 0.8°C since 1870, with the last decade being the warmest on record. Direct measurements (thermometers) have been taken over a large part of the Earth since 1870. Indirect measurements (ice cores, tree rings, coral, etc.) indicate that the last decade has been the warmest for at least a thousand years (Climate Action Network France, 2011). In Africa, the climate is changing and rainfall patterns are expected to evolve across the continent. In many regions, droughts will become more frequent, more intense and last longer. In others, new rainfall patterns will cause flooding and soil erosion. Climate change is emerging as one of the main threats to the continent's development (CTA, 2015). In this context, droughts and rainfall patterns will affect agriculture, which is an important sector of human activity. In the land pacts, climate change is already being observed in the form of drought, flooding and rising daily temperatures. All these climatic indicators are having an impact on agriculture in the study area, and appropriate measures need to be taken to deal with any adverse effects or consequences. Farmers around the world will have to produce more food and other agricultural products on less land, with fewer pesticides and fertilisers, less water and lower greenhouse gas emissions. It will also have to be viable, in other words, sustainable. In the Tèwi basin (Dassa-Zoumè), communities are unanimous on the fact that climate risks have become very noticeable and are disrupting farming activities in particular. These risks include late onset of rain, flooding, pockets of drought in the rainy season, early onset of the short rainy season, poor spatial distribution of rain, early end of rainy seasons, rising temperatures, heavy rain and strong winds (IDID, 2015). Faced with this situation, the development of resilient agriculture will require the use of technologies and practices based on agro-ecological knowledge that enable small-scale producers to counter environmental degradation and climate change in ways that maintain sustainable agricultural growth (CTA, 2014).

Maps representing the present highlight the problems facing communities, including resource depletion, conflict and poverty. They represent the future, reflecting the hopes and dreams of the

4

community, and serve to encourage its members to plan for and achieve positive change (IFAD, 2009). This tool for recording extreme events will enable us to gain a better understanding of the situation and identify the adaptation practices and techniques adopted by the farming population in order to withstand climate change in the land pacts. It is these findings that have led to the study of agricultural production practices in the face of climate change and their geographical location within the land pact area. To do this, we need to analyse :

- What are the effects and consequences of climate change on the land pacts?
- What risks do land pact producers face?
- How producers in the land pacts region are adapting to change

How can we reduce the impact of climate change on the sector in order to maintain the current local dynamics?

To better understand these questions, hypotheses and objectives are put forward.

1.1.2. Hypotheses and research objectives

1.1.2.1. Research hypotheses

✓ The rise in temperature and rainfall are indicators of climate change that influence agricultural production in the Bantè-Savalou area.

✓ certain practices adopted are appropriate in the face of climate change in the Bantè-Savalou area.

1.1.2.2. Research objectives

The main aim of this study is to help improve the conditions under which farmers can adapt to climate change by selecting the most suitable practices in the land pact area.

Specifically, the aim is to analyse :

❖ indicators of climate change on agricultural production in the land pact area.

❖ agricultural practices adopted in the face of climate change in the land pact area.

1.1.3. Literature review

1.1.3.1. Clarification of concepts

In order to understand this study, it is important to define the key concepts precisely so as to eliminate any common-sense confusion.

➢ **Land of land pacts**

The concept of the land pact country originated in the cooperation between the communes of Savalou and Bantè. It is an alliance established between villages to serve as non-aggression or mutual assistance agreements. Article 176 of Title VI, entitled DE LA COOPERATION INTER-COMMUNALE, states: "Several communes may decide to join forces to build and manage facilities and create services of inter-communal interest and utility. In this case, an agreement determines the rights and obligations of each of the parties". These are the terms on which the country of land pacts was created. Benin has 24 shared development areas (DAT, 2014).

➢ **Territory**

Territory in fact refers to the existence of the State, whose legitimacy is largely measured by its ability to guarantee territorial integrity (Brunet et *al.* 1992). For C. Charbonnier (lexique de Géographie), it is a delimited space to which the inhabitants have the feeling of belonging and which they have appropriated through their habitat, their developments, their activities, etc. In other words, it is an area of land occupied by a human group or which depends on an authority (State, province, town, jurisdiction, local authority, etc.).

➢ **Climate change**

Climate change is defined as a variation in the state of the climate that can be detected by changes in the mean and/or variability of its properties and that persists for an extended period, typically decades or longer (IPCC, 2014). In this research, climate change corresponds to a variation in "mean time" observed in a given region over at least 30 years. Mean weather includes all the elements that we usually associate with the weather, namely temperature, wind characteristics and precipitation.

➢ **Cartography**

"Cartography is the totality of scientific, artistic and technical studies and operations, based on the results of direct operations or the exploitation of documentation, with a view to the preparation and establishment of maps, plans and other forms of expression, as well as their use". (UNESCO, 1966). This makes it easier to reconstruct information on the territory of the land pact country.

➢ **Agricultural production**

It can be defined as an economic activity combining factors and materials. Its purpose is to transform and develop the rural environment in order to obtain plant and animal products that are useful to man and satisfy the needs of society (Hachette, 1997). It is also the set of activities developed by man, the purpose of which is to transform his natural environment in order to produce the plants and animals that are useful to him, in particular those needed for his food (Services Québec, 2017).

All these definitions are included in the present work.

1.1.3.2. What we know

Numerous studies have been carried out to gain a deeper understanding of cartography and climate change and, above all, to demonstrate the decisive role of cartography in development. The lesson that emerges is that the climate change controversy is not just an environmental and economic issue, but above all a human rights issue (Haines and Reichman, 2008).

Differences in perception of climate change have led to variability in the forms of adaptation to minimise the effects of climate change. Adaptation by rural populations is a critical aspect in developing countries, where vulnerability is high because of the limited resources of local communities. Adaptation is a kind of ecological, social or economic adjustment in response to observed or future climate change, in order to mitigate its impacts (IPCC, 2001; Adger *et al.* 2005). These adaptations are noted at both individual and community level and are motivated by a multitude of factors such as trade (Smit *et al.* 2000), social networks (Adger, 2005) or by individual actions, especially at the level of agricultural practices, in the case of rural populations. Adaptation could therefore help populations to guarantee their food and income and secure their well-being in the current context of climate change and socio-economic conditions reflected in climate variations, droughts or floods (Kandlinkar and Risbey, 2000).

Agriculture has a special place as a sector that is highly sensitive to climatic hazards. It is therefore important for those involved in agriculture to be fully aware of the effects of climate change, both now and in the future (Part 1 - Agriculture and adaptation to climate change). Climate change will therefore have an impact on the biotechnical component of production. The increase in carbon dioxide and other greenhouse gases in the atmosphere, the rise in temperature, changes in rainfall patterns, and therefore in the various water balance terms (evaporation, drainage, run-off), changes in cloud cover, and therefore in the radiation balance: all the bioclimatic factors that govern the functioning of ecosystems are set to change, and it is therefore necessary first and foremost to predict and quantify these changes and their consequences (Seguin, 2010).

Faced with climate change, its effects on ecosystems and people's adaptations, there is an urgent need in Africa to better understand and characterise these climatic variations and, above all, to analyse local perceptions on the subject. Characterising the adaptations of local populations in the context of climate change will enable us to better analyse, understand and disseminate the best adaptations, taking account of specific regional climatic features. This needs to be combined with yield improvements based on biotechnology and involving better photosynthesis, nitrogen uptake, resistance to drought and other impacts of climate change (Conway, 2012). Similarly, the use of cartographic-based maps is very important, particularly participatory maps, which provide a reliable visual representation of a community's perception of where it lives and its key features. These include the representation of physical features and natural resources as well as the socio-cultural characteristics known to the community. Participatory mapping is multidisciplinary. What fundamentally distinguishes it from traditional mapping and map-making methods is the process by which the maps are created and the uses to which they are put (IFAD, 2009). From a purely rational point of view, adaptation and mitigation should be considered as two sides of the same coin, given their complementary nature (Stehr and von Storch, 2008). Indeed, the inertia of the climate system is such that the various adverse effects of global warming will be inevitable throughout the 21ème century (Meehl et *al*, 2007). The decision-making processes involved in adaptation and the role of the players involved have therefore hardly ever been studied (Dovers and Hezri, 2010).

1.1.4. Reasons for choice

This research project, entitled *"Mapping agricultural production practices in the face of climate change in the land pact area (Savalou-Bantè)"*, falls within the geomatics discipline, which is nothing other than an in-depth analysis of the management of the geographical and agronomic environment, and specifically in the field of precision agriculture in geomatics. The choice of the communes of Savalou and Bantè (land pact countries) is based on the following factors:

- ❖ Espace de Développement Partagé (shared development area) created by the Délégation à l'Aménagement de Territoire (DAT, 2014) and named "pays des pactes des terres" (land pact country) brings together the communes of Savalou and Bantè.
- ❖ The lack of formalisation of this development area, despite its socio-cultural homogeneity and open and remarkable cooperation.
- ❖ An area of cross-border dynamics to be developed in central Benin.

By selecting these factors, it is possible to understand the potential of this shared development area and to see its cross-border dynamics. To do this, we need a suitable methodology, which is presented in several stages below.

1.2. Methodological approach

1.2.1. Equipment

1.2.1.1. Laboratory equipment

The GIS equipment used is a computer, GIS software (Arc GIS, QGIS, ENVI, etc.) and a vector database from the following sources: IFN-CENATEL, 2006 LANDSAT ETM 7+ satellite image, Sokodé sheet. This vector database is supplemented by image data (raster model) and digitised as required. GPS and cameras are used to collect the data.

1.2.1.2. Field equipment

Field equipment included a GPS, a camera and the 1/600000° general map of Benin, IGN-1992.

1.2.2. Data collection method

1.2.3. Literature search

It consisted of researching general and scientific literature. A large number of documents were consulted, and the types of information gathered varied (Table I).

Table I: Documentation centres visited and types of information collected

Centre	Type of documents	Types of information
FASHS	Memoirs	General and methodological information
UAC	Books, reports and articles	General and methodological information
INSAE	Database	General information on socio-demographic variables
ASECNA LARES	Database Books, theses, dissertations, articles, reports	Climate data General information
Internet	Books, theses, dissertations, articles, reports	General information

Source: Field data, 2017

The table shows the various sources of the documents we used and the types of information.

1.2.4. Preparatory work

Non-probability quota sampling was used (table II). Sampling covered two (02) arrondissements in each commune, i.e. a rate of 30 people surveyed per arrondissement. The choice of arrondissements was based on the predominance of ethnic groups in each commune. Two different ethnic groups were chosen in each commune (table II).

Table II: Bantè and Savalou districts.

Communes	Administrative divisions	Number of households	Total	Male	Female	Size household	Number of people surveyed
Bantè	AKPASSI	2215	12967	6407	6560	5,9	30

	DISTRICT						
	ARRONDISSEME NT BANTE	3465	17682	8588	9094	5,1	30
Savalou	TCHETTI DISTRICT	2032	11897	5848	6049	5,9	30
	LAHOTAN DISTRICT	1290	7455	3606	3849	5,8	30

Source: INSAE-RGPH-4 (2013)

The table shows the arrondissements and the number of people surveyed in each commune.

1.2.5. Field work

In-depth individual and group interviews, the semi-structured interviews carried out as part of this research enabled us to gain a better understanding of the subject; to gain a better understanding of the various crop and livestock production practices; and finally, to make the link between practices and climate change.

Focus group discussions were then held to complement the in-depth individual interviews, in order to gain a better understanding of the opinions, motivations and behaviour of the strategic players involved in the farms. Direct observation made it possible to cross-reference the information collected with the findings in the field in order to assess the data collected.

1.2.6. Data processing and results analysis

Data processing and analysis of the results relate to the nature of the information gathered in the documents and in the field. It consists of manually analysing the answers to the questions, followed by the actual processing. The data collected was processed using IBM SPSS Statistics 21, Microsoft Excel 2013 and Word 2013 for formatting tables and graphs and for analysis. This data was used to produce maps of the physical features of the area and then supplemented with field data to produce thematic maps of the territory of the land pact country.

CHAPTER 2

PRESENTATION OF THE STUDY AREA

Chapter II presents the study area and its geographical location, followed by physical and human features and the activities carried out on the land pacts.

2.1. Geographical and administrative location of the study area

Located in West Africa, Benin is a country subdivided into 77 communes and 12 departments. The land pacts are located in the central regional development pole, in the collines department, covering an area of 5,349 km² (INSAE, 2002).

The département des collines has 06 communes divided into 03 shared development areas. The territory of the Pays des Pactes de Terre is located between latitudes 7°30' and 8°40' North on the one hand, and longitudes 1°30' and 2°10' East on the other. It comprises two (02) communes, 23 arrondissements and 111 villages.

It is bordered to the north-east by the department of Borgou, to the north-west by the department of Donga, to the east by the communes of Glazoué and Dassa-Zoumé, to the west by the Republic of Togo and to the south by the department of Zou (Maps 1 and 2).

Map 1: Geographical location of the land pacts territory

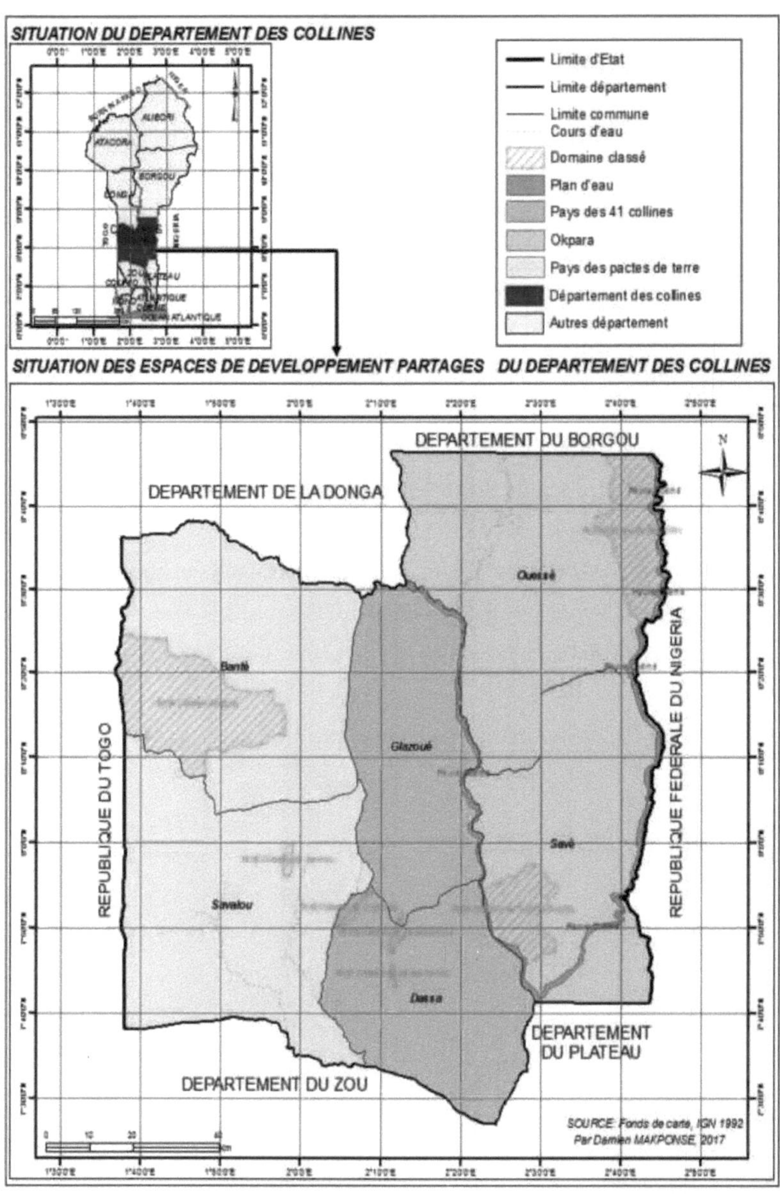

Analysis of Map 1 shows that the land pacts area is located in the hills department and lies to the west of two development areas, Okpara and the 41 hills country. It also borders the Republic of Togo.

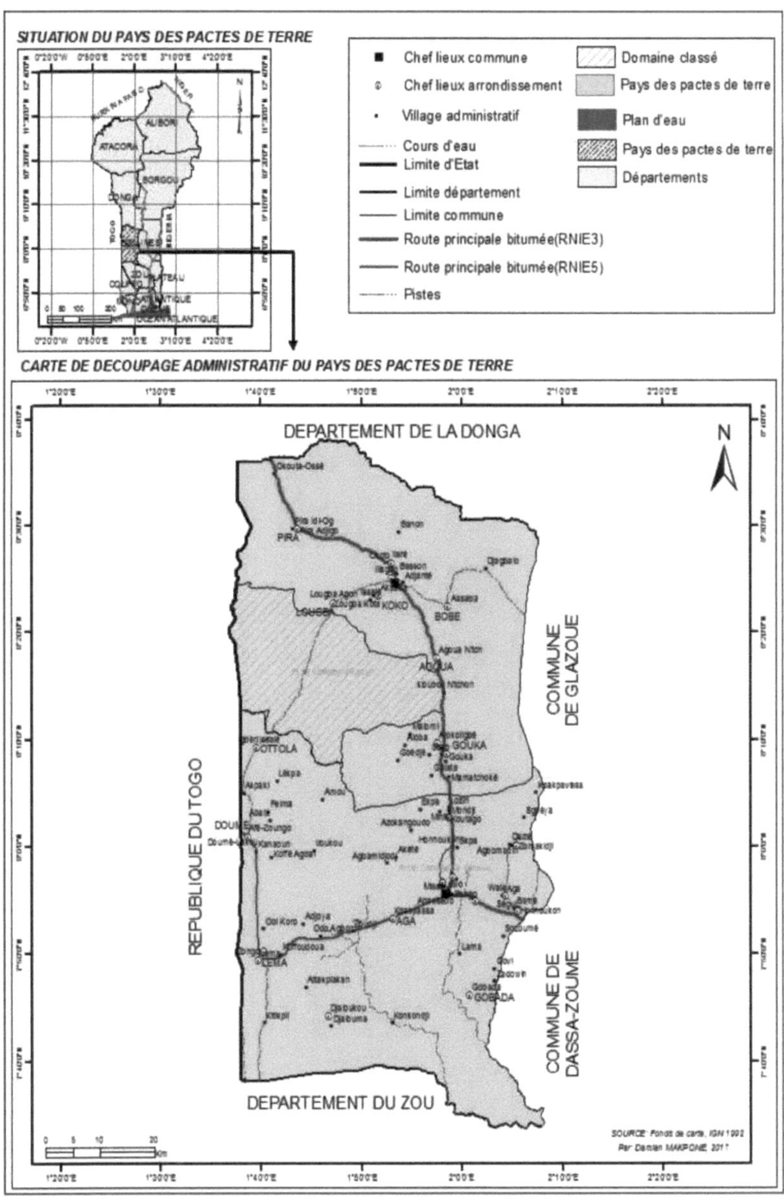

Analysis of Map 2 shows that the land pact territory is made up of two communes, 23 arrondissements and 111 administrative villages. It also includes two classified forests: the Agoua classified forest and the Savalou classified forest.

2.2. Physical features

❖ **Relief**

The land of the land pacts has a rugged relief. The region contains some of the country's rocky massifs. Indeed, the commune of Savalou is characterised by hilly outcrops with a north-south axis on a crystalline peneplain resting on Precambrian material of the old granite-gneissic basement.

It rises to an altitude of between 120 and 500 metres, with a central range peaking at almost 500 metres and stretching for some twenty kilometres between the urban centre and the district of Kpataba, giving the commune its name of "hill country".

A little further afield, the Commune of Bantè is part of the Benin plain, a crystalline peneplain with altitudes ranging from 200m in the south (Gouka) to 300m in the north towards Pira, with peaks of over 400m.

In terms of hydrography, the land of the land pacts is watered by seasonal rivers, the main ones being : Agbado, Klou, Gbogui, Azokan and Zou, which are around 161km long (PDC2). The relief has a negative impact on agricultural production in the sense that the high-latitude landscape forces producers to adopt specific behaviours on the one hand, and on the other, it allows the low-latitude landscape to benefit from alluvial deposits coming from the top of the hills for crop development. All this is shown on Map 3.

Map 3: Relief map and hydrography of the land pacts territory

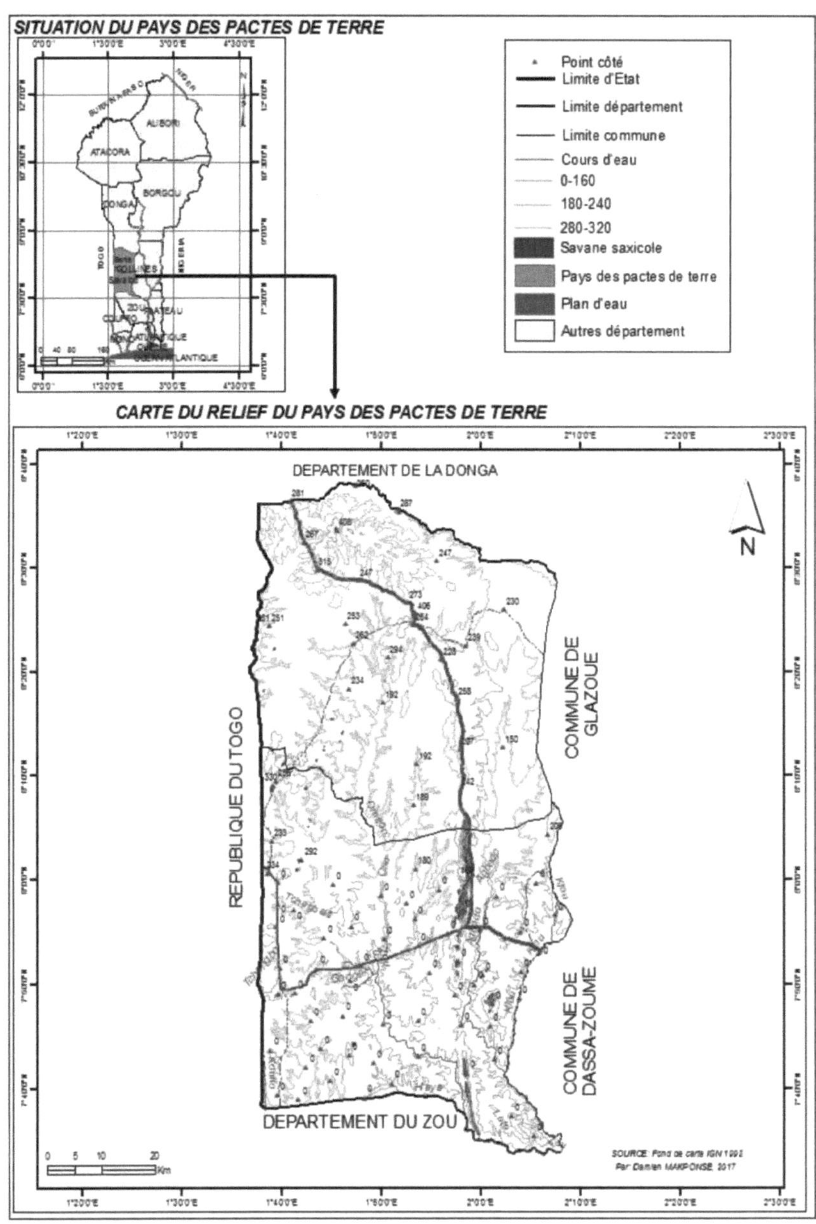

Analysis of Map 3 shows the relief of the area, which is somewhat uneven, with sides ranging from 0 m to 408 m in height. It also has a relatively undeveloped hydrography, with the presence of a few major rivers such as Agbado, Klou, Gbogui, Azokan and Zou.

❖ **Climate**

The territory of the land pacts belongs entirely to the Sudano-Guinean climate zone. But despite this, there is diversity in terms of rainy seasons.

Savalou has a Sudano-Guinean climate, with two rainy seasons (March to July and September to November) and two dry seasons (December to March and August).

Average rainfall is 1,150 mm. Temperatures are high throughout the year, with minimums between 23 and 24°C and maximums between 35 and 36°C (PDC2). The commune of Bantè enjoys a Sudano-Guinean climate, with a dry season from December to March, and a rainy season from April to November.

The highest temperatures are recorded in February, when they exceed 37°C, while the lowest temperatures are recorded in September, when they hover around 23°C. The average rainfall is 1095 mm (PDC2).

The diversity of rainy seasons in each commune raises questions about agricultural production and farming practices.

❖ **Land use**

The territory of the land pacts is characterised by several vegetation formations. The vegetation is of the gallery type with a small dense forest followed by open wooded savannah forests, plantations, rocky slab, crop and fallow mosaics, saxicolous savannahs and settlements.

The dominant plant formation is a shrubby, saxicolous savannah. This sector is mainly made up of granite and gneiss. This constitutes an obstacle to agriculture because of the rocky slab and granites (Map 4).

Map 4: Land use map of the land pacts area

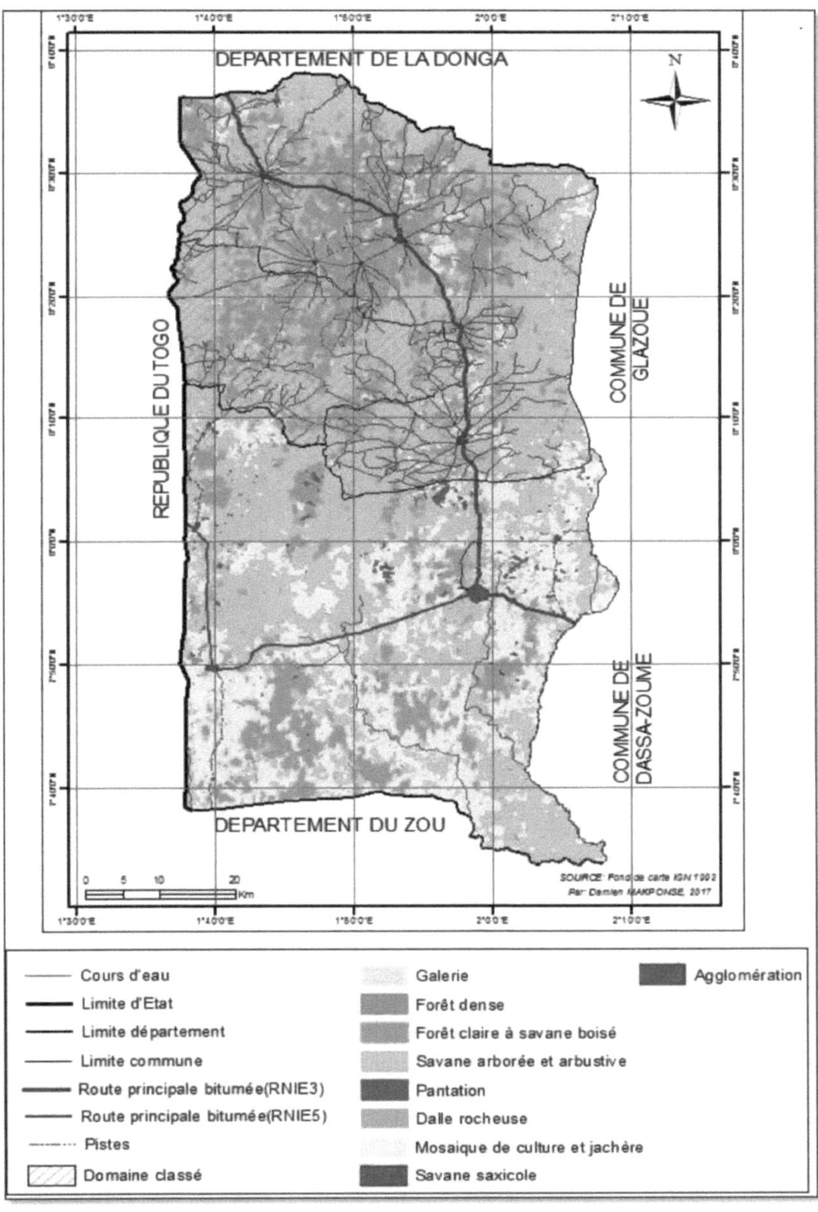

Analysis of map 4 shows the vegetation of the area, with galleries, dense forest, open forest with wooded savannah, tree and shrub savannah, plantations, rocky slab and saxicolous savannah. The vegetation is dominated by savannahs, mosaics of crops and fallow land.

2.3. Human traits

✓ **Demographics**

The territory of the land pacts has 251730 inhabitants including 122805 men and 128925 women according to RGPH 2013(INSAE, 2016). According to the same source in 2017, this population is estimated at 282497 inhabitants with an average growth rate of 2.65% for the territory and a density of 47 inhabitants/km^2 . The urban population was 5,546 in 2013. The most populous commune is Savalou (162216hbts) followed by Bantè (120281hbts) according to the 2017 INSAE projection. This high population availability provides a large workforce for agricultural production in the sector.

2.4. Activities

With 420,980 hectares of arable land available for cultivation, the territory of the land pacts is highly sought-after by agricultural settlers. The economic potential of the commune of Savalou is fundamentally based on agriculture in both communes. Agriculture is extensive in the commune of Savalou and subsistence in the commune of Bantè. Food crops include maize, cassava, yams, cowpeas, voandzou, rice (produced on a small scale in the lowlands), market garden produce (tomatoes, peppers, okra, courgette seeds) and cash crops (groundnuts, soya, cotton, cashew nuts).

Trade is mainly carried out by women around the large markets. They market agricultural products and by-products and resell manufactured goods, often in retail form.

Industrial activity is still in its infancy, with the gradual development of small-scale wood and agricultural product processing units (Monographie des deux communes: April, 2006).

✓ **Other socio-cultural characteristics**

The study area has a peaceful socio-cultural context, with traditional chieftaincy as part of its cultural heritage. In terms of associations, there is a certain dynamism that favours cultural (development associations) and economic (women's, men's or mixed producers' groups) groupings. Several religions coexist in the region, the most important of which are the revealed religions (Christianity and Islam) and the traditional religions commonly known as animism. One of the most respected practices is the belief in the "land pact", which is an alliance established between villages to serve as a non-aggression or mutual assistance agreement.

CHAPTER 3

AGRICULTURAL PRODUCTION AND CLIMATE CHANGE INDICATORS IN THE LAND PACT REGION

This chapter deals with the issues of agricultural production and climate change indicators in the area covered by the land pacts.

3.1. Socio-economic characteristics of farmers.

The characterisation of the respondents helps to understand and link the various factors that explain the dynamics in the area. The sample has an average age of 38: the oldest farmer is 68, the youngest is 23, and the majority of respondents are between 35 and 45. In terms of level of education, 11% of respondents had attended secondary school, 20% primary school and over 60% had no education.

In the two communes surveyed, the economic activity of the interviewees is dominated by agricultural production (60%), and the main income-generating activity is trade in agricultural products. Agricultural activities are carried out collectively by men and women. They practise rain-fed agriculture.

3.2. Farm characteristics

The farms surveyed in the land pacts area are small; 60% of farmers have small scattered areas of no more than 3 ha, while those with medium-sized areas account for 30% (between 5 and 7 ha) and large areas account for 10% (more than 10 ha). The small areas are predominantly used by farmers growing produce for consumption. Large areas are used mainly for crops for consumption and cash crops such as cotton, mahogany, etc.

In the land pact area, agricultural production is essentially rain-fed and depends on the two (02) rainy seasons that prevail there. Thus, the level of production obtained at the end of a year (agricultural season) is often determined by the behaviour of the rains. According to APRM-2016 data, the main crops grown in the Savalou commune are maize, cotton, cassava and yams, while the main crops grown in the Bantè commune are maize, cassava, yams and rice. The cross-tabulation between the two communes also shows that cotton production is higher in the commune of Savalou, whereas in the commune of Bantè, cotton has been abandoned in favour of rice (Photo 1).

Photo 1: Yam and rice field at Lahotan

Shot: MAKPONSE, July 2017

Plate 1 shows a field of yams and rice on rich, fertiliser-free land in the land pact area.

3.3. Cultivation techniques

In the Pactes de Terre region, the farming techniques used are: the use of fallow land to replenish the soil's nutrients naturally; the integration of livestock farming with cultivation on fallow fields; the combination of several crops on the same ridge to ensure the stability of the ridges and good soil management; the retention of crop residues on the fields; the size and layout of the ridges and rows on cultivated plots. These techniques have evolved over time in response to the effects of climate change.

3.4. Trends in agricultural production

> **Growth in crop production**

The trend in production in the land pact area has been up and down. Over the period from 2010 to 2016, the main crops are cassava, maize, cotton, rice and yams. The area covered by the land pacts produces 20% of the cassava, 25% of the maize, 11% of the yams, 21% of the rice and 56% of the cotton in the hill department, as shown in Figure 1.

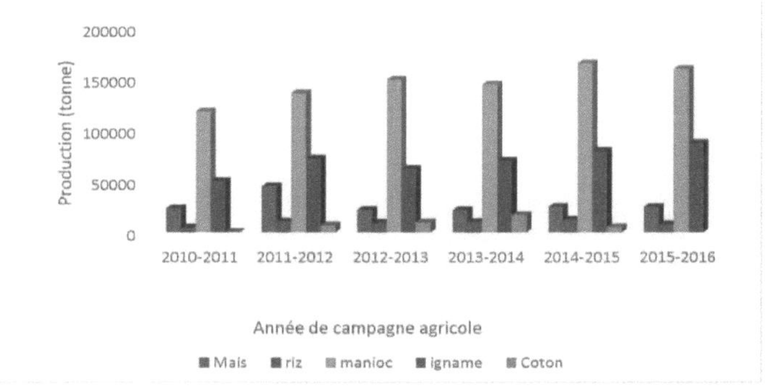

Figure 1: Change in agricultural production on land pacts (2011 - 2016)

20

Analysis of the data for each commune shows that production in the commune of Bantè has moved up and down. The main crops are maize, rice, yams and cotton. Over the period from 2010 to 2016, production of cassava, yams, maize and cotton has followed the same pattern, as shown in Figure 2.

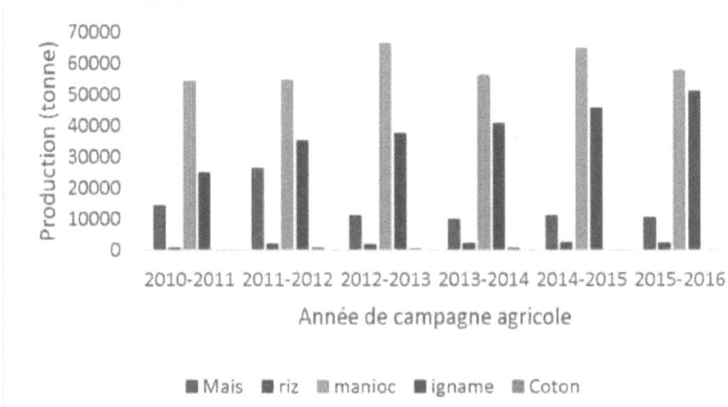

Figure 2: Changes in Bantè's common agricultural production (2011- 2016)

In the commune of Savalou, production has moved up and down. The main crops are maize, rice, cassava, yams and cotton. Over the period from 2010 to 2016, production trends were as follows: cassava, yams, maize, rice and cotton, as shown in Figure 3.

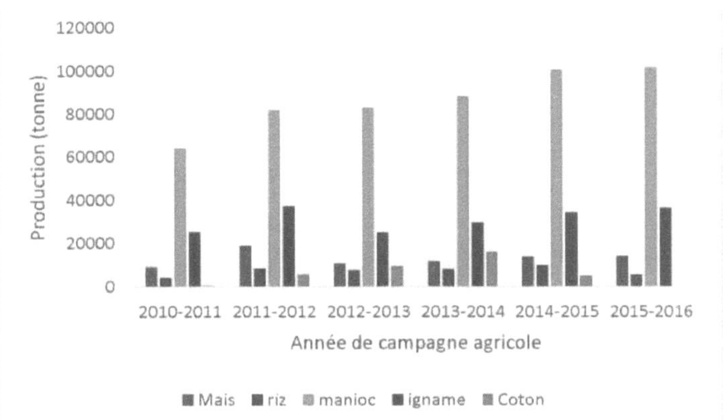

Figure 3: Changes in Savalou's common agricultural production (2011- 2016)

➢ **Yield trends**

In terms of yield, production increased rapidly, peaking in 2012 before gradually declining in subsequent years until 2013. From 2013 to 2016, it remained constant. This raises the question of what is causing the drop in yield, as shown in Figure 4.

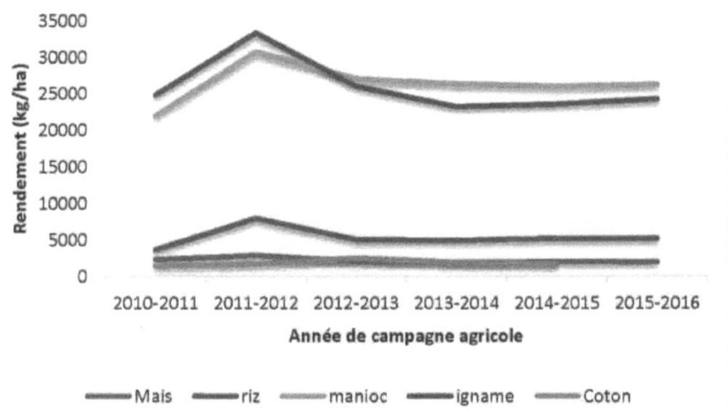

Figure 4: Yield trends in the land pacts territory (2011 -2016)

Source: MAEP-2016

An analysis of each commune shows that in the commune of Bantè, production yields for cotton, yams, cassava and rice peaked in 2011-2012 before gradually falling in 2013. From 2013 to 2016, it remained constant. Maize, on the other hand, remained constant throughout the period. This raises the question of what is behind the drop in yield, as shown in Figure 5.

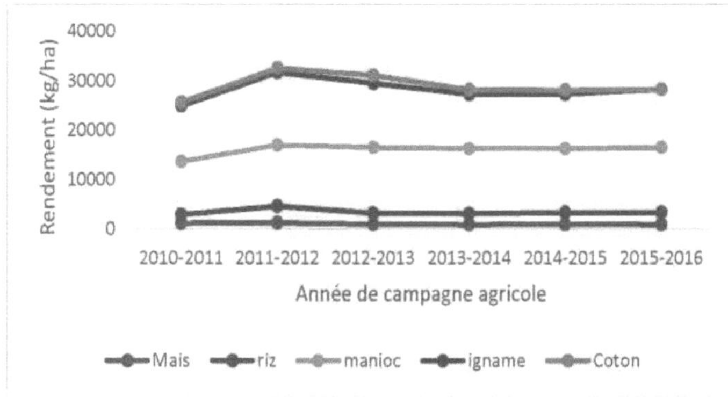

Figure 5: Change in yield in the commune of Bantè (2011-2016)

Source: MAEP-2016

In the commune of Savalou, production yields for yams, cassava, rice and maize peaked in 2011-2012 before gradually falling in 2013. From 2013 to 2016, yields remained constant. Cotton, on the other hand, remained constant throughout the period. This raises the question of what is behind the drop in yield, as shown in Figure 6.

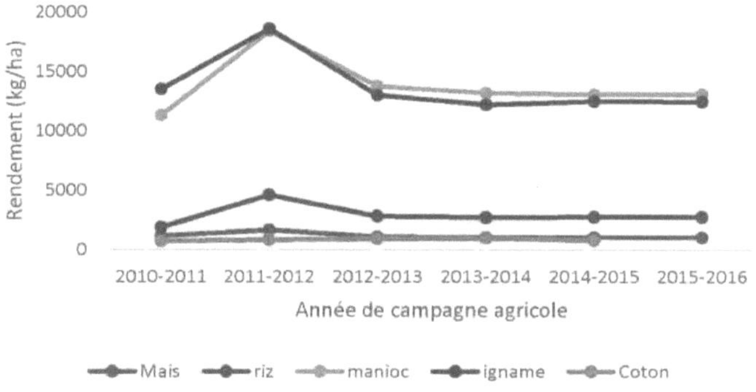

Figure 6: **Changes in yield in the commune of Savalou (2011-2016)**

Source: MAEP-2016

3.5. Climate change

3.5.1. Temperature

Considering the data from 1986 to 2016 on the territory of the land pacts, the average quarterly temperature in January is 28.1°c whereas the normal is 29.6°c, i.e. a decrease of 1.5°c; that of April is 28.6°c whereas the normal is 28°c, i.e. an increase of 0.6°c; August's temperature was 25°c compared with the normal 25°c, i.e. a constant for the month of August; October's temperature was 26.6°c compared with the normal 27°c, i.e. a decrease of 0.4°c over time. From this analysis, it can be seen that the average temperature in the land pact area changed gradually from January to April, when it peaked, before decreasing until October, as shown in Figure 7. This change in temperature will have an impact on agricultural production.

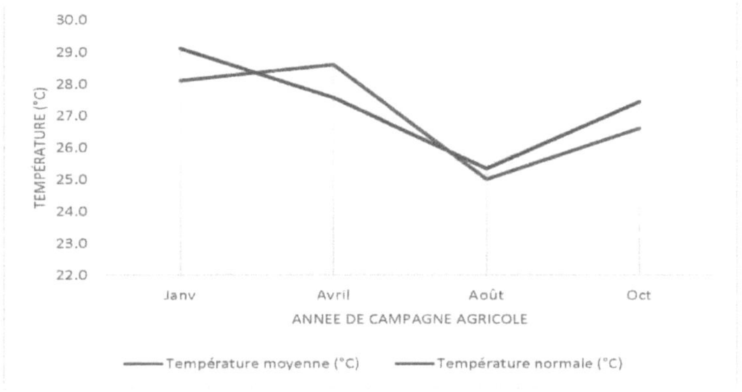

Figure 7: **Temperature trends in land pacts (1986-2016)**

Source: météorologie-Bénin, 2017

23

3.5.2. Rainfall

In the land pact area, quarterly rainfall was 4.5 mm in January, compared with a normal 30 mm, a drop of 25.5 mm; April rainfall was 112.5 mm, compared with a normal 134 mm, a drop of 21.5 mm; August rainfall was 142 mm, compared with a normal 162 mm, a drop of 20 mm; October rainfall was 113.8 mm, compared with a normal 43 mm, a drop of 20 mm; August's rainfall was 142 mm, compared with the normal 162 mm, i.e. a decrease of 20 mm; October's rainfall was 113.8 mm, compared with the normal 43 mm, i.e. an increase of 70.8 mm.

This analysis shows that rainfall is lower than normal from January to August, before rising above normal from mid-August to October, as shown in Figure 8. This will have an impact on agricultural production in the region.

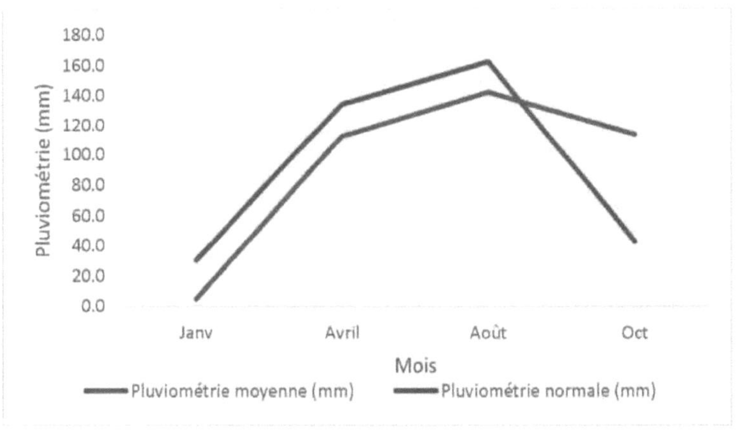

Figure 8: Rainfall trends in land pacts (1986-2016)

Source: météorologie-Bénin, 2017

3.5.3. Evapotranspiration (ETP)

The average evapotranspiration in January was 28.1 mm compared with a normal of 132.3 mm, a decrease of 104.2 mm; the average evapotranspiration in April was 28.6 mm compared with a normal of 133 mm, a decrease of 104.4 mm; the average evapotranspiration in August was 25 mm compared with a normal of 106.3 mm, a decrease of 81.3 mm; August's rainfall was 25 mm, compared with the normal 106.3 mm, a decrease of 81.3 mm; and October's rainfall was 26.6 mm, compared with the normal 122 mm, a decrease of 95.4 mm.

This shows a decrease in evapotranspiration throughout the year, as shown in Figure 13. This will have an impact on the soil and agricultural production.

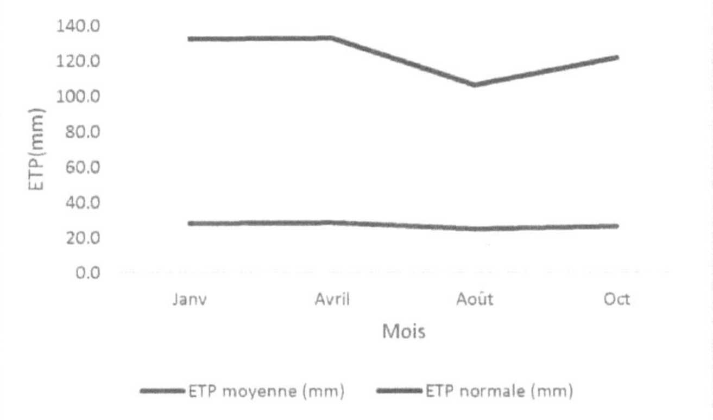

Figure 9: **Change in ETP for land pacts (1986-2016)**

Source: météorologie-Bénin, 2017

3.6. Climate change indicators for the land pact region

An indicator is "data selected from a larger statistical set because it has a particular significance and representativeness". As its name suggests, it is information that *indicates* something. It simplifies information to shed light on sometimes complex phenomena.

In the land pacts, the indicators of climate change observed are : a delay in the onset of rains, observed by 92.2% of respondents, the appearance of pockets of drought in the rainy season, the early cessation of the rainy season, the increase in daily temperature, the appearance of violent winds, observed by 45.8% of respondents, the drying up of rivers and bodies of water during the dry season, the multiplication of pockets of drought during the rainy season, the increase in the level of plant attack and, finally, the poor spatial distribution of rain is observed throughout the territory. Violent winds are more likely to occur in the commune of Savalou than in Bantè, because of the vegetation, which is even more prevalent in Bantè.

CHAPTER 4

RISKS AND METHODS OF ADAPTING TO CLIMATE CHANGE

This chapter looks at the climate risks in the land pacts and the adaptation methods adopted by the farming population.

4.1. Main climate risks in the land pact area

Benin, like most African countries, is exposed to disasters caused by climatic risks, in particular extreme drought, strong winds, late and intense rainfall, flooding and excessive heat.

Climate risk is a risk associated with increased vulnerability to variations in climatic indices such as temperature, wind, snow or precipitation.

The climate risks observed by the population include: unusual flooding, pockets of drought, early drought, rising temperatures, violent winds and poor spatial distribution of rainfall. Of these risks, 83% of respondents said that there were pockets of drought, 71% said that there was a poor spatial distribution of rainfall, and 39% said that there was a rise in temperature throughout the country, as shown on Map 5.

Map 5: Main climate risks for land pacts

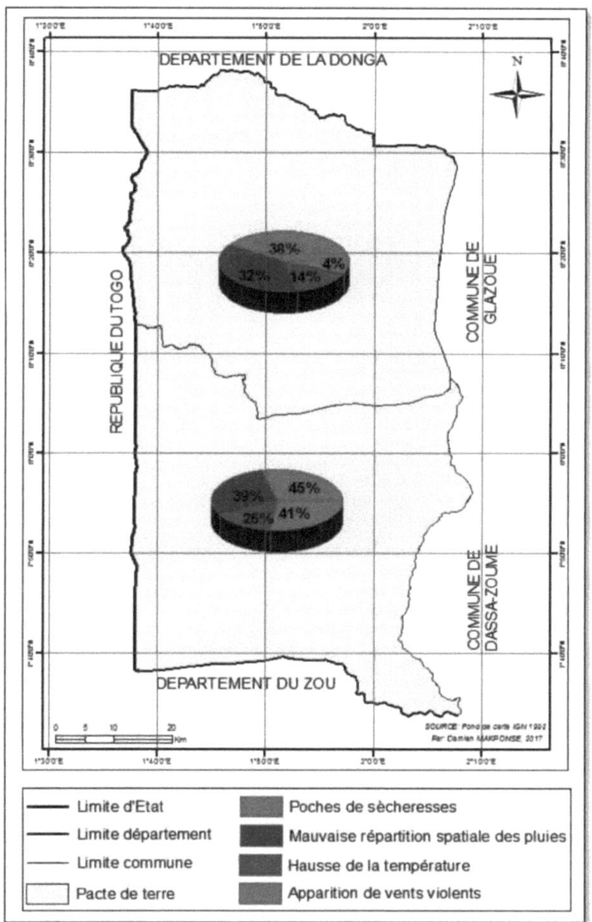

Map 5 shows that the main climatic risks are drought, poor spatial distribution of rainfall and rising temperatures.

4.2. Physical impact of climate hazards on land pacts

In Benin, the land pacts have a number of negative impacts, including lower yields, low seedling emergence rates, delayed plant growth, plants unable to complete their vegetative cycle, plant lodging, crop flooding, pest resistance to plant protection treatments, total loss of production, high post-harvest losses and destruction of fields by animals. Plant lodging and pest resistance to phytosanitary treatments are more noticeable in the commune of Savalou than in Bantè (Map 6).

Map 6: Physical impacts of climate change on the land pacts territory

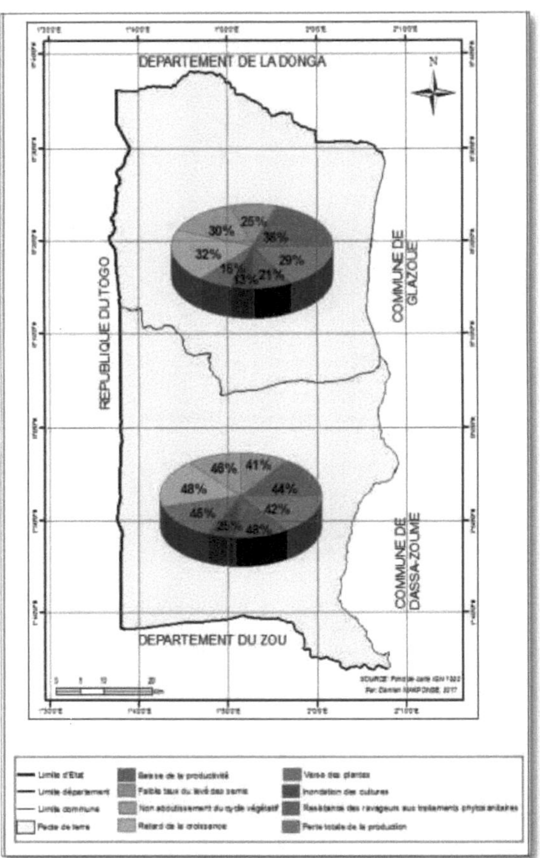

Map 6 shows the various impacts on agricultural production in the land pacts area. The most frequent impacts are: reduced productivity, plant worms, total loss of production, low rate of seedling emergence and stunted growth.

4.3. Methods for adapting land pacts to local conditions

The Pactes de Terre region is faced with a number of climatic risks, including late onset of the rains, the appearance of pockets of drought during the rainy season, poor spatial distribution of the rains and early cessation of the rains. These risks expose the population to a number of difficulties to which adaptations and solutions must be found. Generally speaking, solutions can be found in techniques and practices such as : abandoning crops or crop varieties; adopting new crops or crop varieties; relocating crops; gradually changing the agricultural calendar and technical itinerary; connecting the ends of ridges and anti-erosion management; mulching soil with plant debris; agro-forestry;

Development of new varieties, including genetically modified organisms (GMOs); use of inputs; exploitation of landscape units; longitudinal trenches to encourage water infiltration on sloping soils; use of Mucuna or Aeschinomenae as cover plants; use of "rainmakers"; use of low-lying areas; late and early planting, etc.

Consideration of the various practices observed in the study area shows that they have evolved to adapt better to climate change. Rainfed agriculture is characterised by climatic precariousness, which forces farmers to implement a number of adaptation strategies in order to manage water, land, crop pests, inputs and so on. According to our field surveys, the level of use and performance of techniques and practices will show which are adapted to climate change. Among these practices, the most widely used are :

> **Use of lowlands**

Farmers are making increasing use of the lowlands. This can even be seen in areas where they were not previously used. Sorghum is being grown in low-lying areas traditionally used for rice cultivation, as flooding is becoming rare. The majority of farmers in land pacts adopt this method, especially at the start of the agricultural season when rainfall is scarce.

> **Abandonment of crops or crop varieties**

Nearly 60% of farmers in the land pact area say they use it, with an average success rate of 93.3%. The level of use is average in Bantè and high in Savalou. It is used to prevent the appearance of pockets of drought in the rainy season by 68.3% of respondents, and to prevent the delayed onset of rains by 28.3% of respondents in the region. In view of climate change, growers are abandoning certain varieties in favour of others.

> **Adoption of new crops or crop varieties**

Growers use it because of the irregularity of the rains, in order to improve yields and have resistant and satisfactory varieties. It is used to combat the appearance of pockets of drought in the rainy season by 59.2% of respondents and the delayed onset of rains by 36.7% of respondents in the region. The level of use is average (80%) and high (15.8%), with an average performance of 79.2% and satisfactory performance for 15.8% of respondents.

> **Gradual change in farming calendar and technical itinerary**

The gradual change in the agricultural calendar and technical itinerary is used to prevent the appearance of pockets of drought in the rainy season by 91.7% of respondents in the land pact area. The level of use is average at 48.3% and high at 22.5%, with satisfactory performance at 65.8% and average at 23.3% of respondents in the area. It takes place when the rains start.

> **Change of plot site**

This practice is observed because of the many changes in rainfall patterns in the area. According to the field surveys, 26.7% of farmers use this method to prevent unusual flooding and 2.5% to prevent rainfall delays. The level of use for changing the location of plots is average, with satisfactory performance.

> **Connecting the ends of the ridges and erosion control measures**

To combat the effects of erosion in the fields, the producers of the land pacts place ridges or anti-erosion structures. Connecting the ends of ridges and anti-erosion structures is used to combat the poor spatial distribution of rain and intense rainfall. The level of use is average, with average performance across the region. This practice enables the population to manage water stress.

> **Mulching the soil with plant debris**

This technique allows them to keep the soil cooler and fertilise it better. Soil mulching is used by 50% of respondents in the region to prevent the rain from starting late. The average level of use is 8.3%, with an average performance of 49.2% of respondents in the region. Farmers often use the debris from soya, maize, sorghum, etc., as a soil conditioner.

> **Use of inputs**

The use of inputs is used to combat delays in the onset of the rains, insect pests and reduced productivity. They are used regularly by farmers in both Savalou and Bantè, with very satisfactory results. Fertilisers stimulate plant growth, but they also affect soil stability. In the area covered by the land pacts, growers use Lambda insecticide for beans, NPK fertiliser for maize, and ADWUMA WURB, CALFOR-G and GLYPHADER herbicides to stop weeds growing. To kill the weeds they use the herbicide KALACH. Despite these advantages, there are a number of disadvantages revealed by land pact producers, such as soil impoverishment over the years, the appearance of unprecedented weeds, etc. (Plate 2).

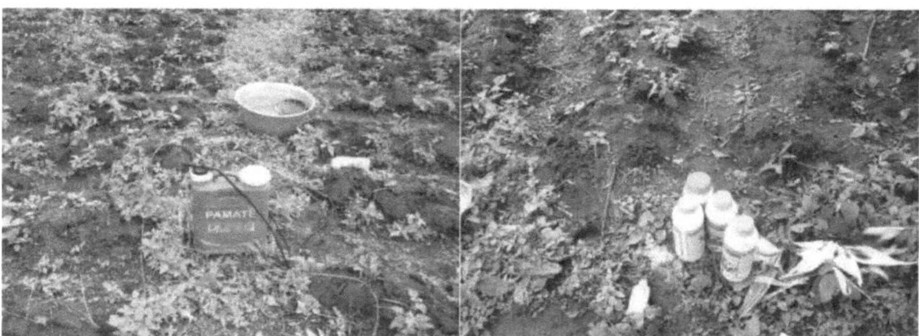

Photo 2: Equipment and inputs used in the fields

Shot: MAKPONSE, July 2017

Photo 2 shows the equipment and inputs used to treat plants and seeds in the fields.

> **Late and early planting**

The delay in the onset of the rains, the rise in temperature and the poor spatial distribution of rain mean that growers have to plant late and early.

In the absence of a water deficit and in an average year, the trials show that for the same sowing and harvesting date, the choice of a later or earlier variety would result in a good yield, according to our field surveys.

Earliness expresses the length of time the plant takes to develop, from sowing to harvesting. For example, the earlier the maize, the faster its cycle and the fewer degree-days (dd) it needs to reach harvest maturity. The level of use is average and somewhat satisfactory.

> **Mound cultivation**

In most African countries where tubers such as yams and sweet potatoes are grown, mounds have been made for generations using the hoe or the daba (a traditional tool of Sahelian farmers, similar to the hoe) to facilitate cultivation. On the land pact territory, this technique is used to facilitate the penetration of water and better evacuate it in case of excess for a good production yield (Photo 3).

Photo 3: Yam cultivation on mounds

Shot: MAKPONSE, July 2017

Photo 3 shows yam cultivation on mounds in the land pact area. The first view shows yam cultivation on simple mounds, while the second view shows yam cultivation on mounds with dry tree support for yam growth.

All these practices and techniques are illustrated in Figure 16.

31

Map 7: Adaptation techniques and practices used in the land pacts area

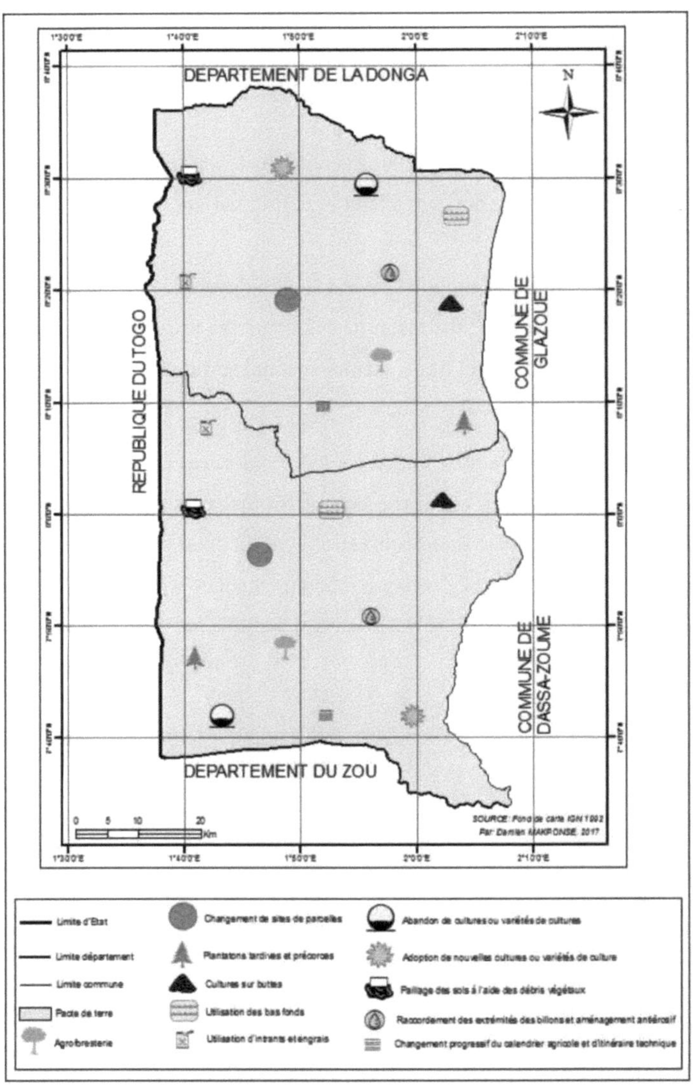

Map 7 shows the different methods of adaptation applied by the farmers in the land pacts. These include: changing plot sites, late and early planting, cultivation on hillocks, use of low-lying areas, use of inputs, gradual change in the agricultural calendar, adoption of new crop varieties, mulching the soil with plant debris, abandonment of cultivation and, lastly, connecting the ends with anti-erosion ridges.

4.4. Suggestions

Climate change in the land pact area has caused a number of problems for agricultural production. Faced with this situation, it is essential to find solutions by making a number of suggestions.

- **For the farming population**
- Raise awareness of bad practices that can exacerbate the effects of climate change, such as the excessive use of agricultural inputs, deforestation, etc.
- Training people in the manufacture and use of organic inputs.
- Raising awareness of the importance of agroforestry.
- **To international agricultural institutions**
- Further research into people's knowledge and practices to limit the effects of impacts, etc.
- Select short-cycle varieties capable of withstanding the effects of climate change and insect pests.
- Develop more efficient growing techniques using fewer agricultural inputs
- Financing irrigation techniques so that they are accessible to producers.
- **In relation to equipment and inputs**
- Reduce the cost of agricultural inputs for farmers.
- Facilitating access to tools and machinery to better manage farming activities.

Conclusion

This study of agricultural production practices in the Pactes de Terre region has enabled us to analyse the farming population's perception of climate change. This analysis enabled us to identify the major risks, the effects and consequences of these risks on agricultural production and, finally, to take account of the existing methods of adaptation in the Land Pacts area.

The field results show that farming households in the land pact area recognise the change in climatic parameters, such as the increase in daily temperature, pockets of drought, poor spatial distribution of rainfall, the delay or sudden cessation of rainfall, and so on. These changes confirm the harmful effects of climate change, such as lower yields, low seedling emergence rates, delayed plant growth, plants unable to complete their vegetative cycle, lodging of plants, flooding of crops, pest resistance to phytosanitary treatments, total loss of production, high post-harvest loss rates and destruction of fields by animals. Over the course of the century, these effects have led people to adopt methods and practices for coping with climate change, such as: improving seeds, improving production techniques, gradually changing the agricultural calendar, using short-cycle varieties and using inputs. These practices, which are used and considered relevant by the population, are not without their faults, as they require support to help the farming population, such as raising awareness of bad practices that can exacerbate the effects of climate change, training the population in the manufacture and use of organic inputs, further research into the knowledge and practices of the population that can limit the effects of the impacts, and so on.

Bibliography

1   ADGER N. W., ARNELL N. W. and TOMPKINS E. L., 2005: Successful adaptation to climate change across scales. Global Environmental Change, 15, 77-86.

2   ASAP, 2015: Programme d'adaptation de l'agriculture paysanne, 11pages.

3   BUSH M. and FLENLEY J., 2007: Tropical rainforest responses to climatic change. Springer, 225 pages.

4   CONSEIL REGIONAL DE PICARDIE, 2011: Plan de Développement Communal de Savalou, Comité de Pilotage de Savalou/ Groupement Intercommunal des collines, 26 pages

5   CONWAY, 2012: How to create resilient agriculture, http://www.scidev.net/en/agriculture- and-environment/opinions/how-to-create-resilient-agriculture-1.html.

6   CTA, 2013: Agricultural resilience to crises and shocks, 63 pages.

7   CTA, 2015: Climate smart agriculture, what impact for Africa, 44 pages.

8   CHARBONIER C. Lexique de géographie, vocabulaire et notions. Collège Pierre Grange, ALBERTVILLE, 36 pages.

9   DAT, 2014: Aménagement du territoire : L'heure du bilan a sonné, 39 pages

10   DAUPHINE et al. 2007: La résilience: un concept pour la gestion des risques, pages115-125.

11   DOVERS, S. R., & HEZRI, A. A., 2010: Institutions and policy processes: the means to the ends of adaptation. WIREs Climate Change, 1: 212-231.

12   DUGUE. M. J., 2012: Caractérisation des stratégies d'adaptation au changement climatique en agriculture paysanne. Capitalization study carried out on AVSF cooperation projects with the support of : Hélène Delille, Sylvain Malgrange, 50 pages.

13 ETS, 2012: A resilient and interconnected city, 6pages.

14 IFAD, 2009: Participatory mapping and good practices, 55pages.

15 FAO, 2010: Climate-smart agriculture, 47 pages.

16 FAO, 2012: Towards the future we want: Ending hunger and making the transition to sustainable food systems, 48 pages

17 GNANGLE et al, 2011 : Tendances climatiques passées, modélisation, perceptions et adaptations locales au Benin pp 27-40.

18 IPCC, 2014: Climate change: indices, adaptation and vulnerability. Edited by Christopher B. Field, Co-Chair of the Working Group of the Department of Global Ecology of the Carnegie Institution of Science and Vicente R. Barros, Co-Chair of Working Group II of the Center for Marine and Atmospheric Research (CIMA), 222 pages.

19 GRUERE G., 2016: Agriculture and water: a real headache, summary No. 43.

20 HAINES and REICHMAN, 2008: The Problem That Is Global Warming: Introduction, Vol. 30, No. 4, pp. 385-393, October 2008.

21 HAMANI, 2007: Adaptation de l'agriculture aux changements climatiques: Cas du département de Téra au Niger, 106 pages.

22 IPCC, 2001a: Climate Change 2001: The scientific basis. Report of IPCC, Working Group I, IPCC, Geneva, http://www.ipcc.ch.

23 IDID, 2015: Rapport d'étude sur les impacts des changements climatiques et avancées en matière d'adaptation: cas du bassin de tèwi dans la commune de Dassa-zoumè, 32 pages.

24 KANDLINKAR M. and RISBEY J., 2000: Agricultural impacts of climate change: If adaptation is the answer, what is the question? Climatic Change, 45, 529-539.

25 LARES, 2000: La problématique de l'intercommunalité dans le fonctionnement des communes béninoise, 260 pages.

26 MARTINAIS, 2008: La cartographie au service de l'action publique, 13 pages.

27 MAEP, 2014: Rapport de Performance du Secteur Agricole, Gestion 2013, 47 pages

28 MEEHL, G.A., et al (2007) Global Climate Projections. In: Climate Change 2007: The Physical Science Basis. Contribution of Working Group I to the Fourth Assessment Report of the Intergovernmental Panel on Climate Change, Cambridge University Press, Cambridge 10 pages.

29 JOHANN DUPUIS and PETER KNOEPFEL, 2011: Les barrières à la mise en œuvre des politiques d'adaptation au changement climatique: le cas de la Suisse, P 188-219.

30 MONTCHO, 2014: Agriculture and climate change in Benin, 4 pages

31 MADIGNIER M. L. et *al*, 2015: Les contributions possibles de l'agriculture et de la forêt à la lutte contre le changement climatique, 83 pages.

32 UNDP, 2012: Rapport de synthèse sur les progrès réalisés en matière de développement durable en Haïti, 68 pages

33 UNEP, 2014: Strengthening the resilience of urban agricultural systems, 47pages.

34 PANA1, 2014: Choosing agricultural technologies for adaptation to climate change in panal intervention communes, 87pages.

35 RAUFFLET E., 2014: De l'acceptabilité sociale au développement local résilient, volume14 N°2 12 pages.

36 RESEAU ACTION CLIMAT-FRANCE, 2011: Changements climatiques, comprendre et réagir, 45pages.

37 SEGUIN, 2010: Impact du changement climatique et adaptation de l'agriculture, 7 pages

38 SALIOU M., 2008: Déterminants du prix de la terre agricole au Bénin, 22 pages.

39 STEHR, N., & VON STORCH, H. (2008): Anpassung und Vermeidung oder von der Illusion der Differenz. GAIA - Ecological Perspectives for Science and Society, 17: 270-273.

40 SMIT B., BURTON I., KLEIN R. J. T. and WANDEL J., 2000: An anatomy of adaptation to climate change and variability. Climatic Change, 45, 223-251.

41 UNION EUROPEENNE/PRODECOM ,2006: Plan de Développement Communal de Bantè, ONG Initiative Développement/Programme Rural d'Appui aux Communes (ID/PRAC), 109 pages.

42 http://www.merid.org/fr

FR/Content/News_Services/Food_Security_and_AgBiotech_News/Articles/2012/mar/21/a.as px/le 30/10/2017 à 11h22mm

Printed by Books on Demand GmbH, Norderstedt / Germany